D0984273

The QM2 Story

CUNARD
Queen Mary 2

The QM2 Story

Chris Frame & Rachelle Cross

The History Press

Also in this series:

The Concorde Story

The Spitfire Story

The Vulcan Story

The Red Arrows Story

The Harrier Story

The Dam Busters Story

The Hurricane Story

The Lifeboat Story

The Tornado Story

The Hercules Story

The QE2 Story

The Titanic Story

Half title page: *The three Cunard Queens,* QM2, QE2 *and* Queen Victoria. *(Cunard Line)*

Title page: QM2 *on a still ocean. (Matt Welsby)*

Published in the United Kingdom in 2010 by
The History Press
The Mill · Brimscombe Port · Stroud · Gloucestershire · GL5 2QG

British Library Cataloguing in Publication Data
A catalogue record for this book is available from the British Library.

Hardback ISBN 978-0-7524-5092-6

In memory of George

Typesetting and origination by The History Press
Printed in Italy

The sheer size of QM2 is overwhelming. (Authors' Collection)

CONTENTS

QM2 in Southampton Water. (Andy Fitzsimmons)

We'd like to send our thanks to everyone who helped us tell *QM2*'s story. Special thanks to...

Commodore R.W. Warwick, Commodore Bernard Warner, Captain Christopher Rynd, Bill Miller, Pamela Conover, Michael Gallagher, Alastair Greener, John Langley and Tim Wilkin for providing their thoughts, information and photographs of *QM2*.

Captain Nick Bates, Chief Engineer Brian Wattling, Food & Beverage Manager Stefan Engl and Marine Supervisor Alan Gould, for their tours and insight into the operations of *QM2*.

Catherine Wood, Christel Hansen, David Hudson, Harriet Johnson, Harry Morley, James Griffiths and Slaven Roje for assisting with factual information and the tour of *QM2*'s bridge.

Everyone at The History Press, in particular Amy Rigg, Emily Locke and Glad Stockdale for their support and hard work.

Alex Lucas, Andrew Sassoli-Walker, Andre van Niekerk, Andy Fitzsimmons, Ben van Zeijl, Colin Hargreaves, Emily Wealleans, Frank Prudent, Fraser McInnes, Holger Jurgensen, Jan Frame, Kyle Johnstone, Marc-Antoine Bombail, Matt Welsby, Pam Massey, Paul A. Tenkotte, Penny Atwell, Richard Edwards, Rob Lightbody, Rob O'Brien, Ron Burchett, Russ Willoughby, Sam Warwick (and the Warwick Family), Scott Ebersold and Thad Constantine for their photographic assistance...

... and our families for supporting us.

A forward view of Cunard's flagship. (Andy Fitzsimmons)

With the introduction of the jet aeroplane in the late 1950s, the passenger ship as a mode of transport turned from a necessity to a novelty. For much of the latter part of the twentieth century, the idea of a modern transatlantic ocean liner was but a faded dream shared by maritime enthusiasts the world over.

Since 1969 Cunard had operated what was thought to be the last of the great transatlantic liners. Their flagship, *Queen Elizabeth 2*, was a unique remnant of a bygone era. Built at a time when the jet had already won the transportation battle, *QE2* maintained a regular Atlantic service for most of her life. She was the sole survivor, a tribute to the past, sailing alone on a vast ocean.

It was felt that when *QE2* eventually did retire, it would see the end of a tradition

A unique liner that fosters the tradition and legacy of British seagoing excellence.

Larry Pimintel, former CEO of Cunard Line.

▲ *QM2's predecessor,* QE2, *is the longest-serving Cunarder in history. (Cunard Line)*

▲ *The magnificent lines of* QM2 *are clearly evident here. (Andy Fitzsimmons)*

that could be traced back to 1840. As such, Cunard invested many millions in maintaining *QE2*, resulting in her enjoying a long and illustrious career. It was felt that the cost of building a successor would be insurmountable.

However, one factor that hadn't been taken into consideration was Carnival Corporation's 1998 acquisition of the Cunard Line. Founded three years after *QE2* entered service, Carnival had grown from a single ship fleet, offering cheap Caribbean cruises, to become the market leader by the time they successfully bid for the veteran Cunard Line.

With Cunard's amalgamation into the Carnival family, the dream of a new breed of ocean liner was re-ignited when Cunard's then CEO, Larry Pimintel, announced their intention to build a 'unique liner that fosters the tradition and legacy of British seagoing excellence'.

This unique liner grew to become the legendary *Queen Mary 2*. She was, at the time, the longest, widest, tallest and most expensive passenger ship ever conceived.

QM2 towers above her escort as she makes her approach to Adelaide. (Ron Burchett)

➤ *The Canyon Ranch Spaclub is one of the finest spas at sea. (Authors' Collection)*

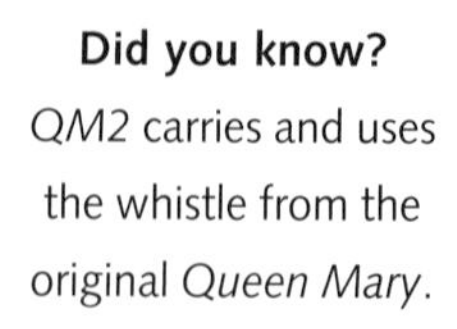

Did you know?

QM2 carries and uses the whistle from the original *Queen Mary*.

Despite her mammoth size, she maintains a timeless elegance that showcases her family connection with the Cunard liners of days gone by.

QM2 is unquestionably one of a kind. From her awe-inspiring size, to her opulent grand duplex suites; her Canyon Ranch Spaclub, to the first planetarium at sea, the ship offers surprising treats for passengers to experience at every turn.

For the next thirty years, and maybe even longer, *QM2* will plough the oceans, a magnificent tribute to that great tradition of ocean travel.

Long may she reign!

Did you know?
QM2 is so large she would not fit into the London Millennium Dome.

Illuminations is the first planetarium at sea. (Authors' Collection)

QM2's 1,132ft length makes her the longest Cunarder of all time. (Matt Welsby)

One of the most important stories in the history of international passenger shipping is that of the Cunard Line. With roots that stretch across three centuries, the story of the Cunard Line started in 1839 when Sir Samuel Cunard successfully won the contract to provide a regular mail service between Great Britain and the United States of America.

In order to fulfill the contract, Cunard was bound to build four new ships. He met legendary naval architect Robert Napier who designed and built his flagship, the RMS *Britannia*. At 207ft long and 34ft wide, Britannia was an impressive size in her day. Built in Greenock, Scotland, she was the first of many Cunarders to be constructed at or near the Scottish town.

Britannia went on to enjoy a profitable career and Cunard's new business flourished. His successes mounted despite growing competition from the American Collins' Line and the British Oceanic Steam Navigation Co. (otherwise known as White Star Line).

However, by the end of the 1800s Cunard Line faced new rivals. The German Norddeutscher Lloyd liner *Kaiser Wilhelm der Grosse* had taken the speed record as the fastest liner in the world. Not only was the new ship extremely fast, she was also larger and more luxurious than her British competitors.

To make matters worse, *Kaiser Wilhelm der Grosse* was not the only German liner introduced that threatened the Cunard fleet. Competition between the two dominant German shipping companies resulted in a battle of one-upmanship, leaving Cunard's fleet a distant third choice for transatlantic travellers.

Sir Samuel Cunard as depicted in QM2's Grand Lobby. (Pam Massey)

RMS Britannia *as depicted in QM2's Grand Lobby. (Pam Massey)*

In response, the Cunard Line commissioned the construction of *Lusitania* and *Mauretania*. These huge ships, each over 700ft long and weighing an impressive 31,000 tons, became known as Cunard's Ocean Greyhounds and soon won the Blue Riband for the fastest transatlantic crossing.

With the pride of the British merchant navy in their fleet, Cunard entered a new phase of growth. *Lusitania* and *Mauretania* were joined in 1914 by the *Aquitania*. Colloquially known as 'Ship Beautiful', *Aquitania* was widely considered the most elegant ship afloat. Although slower than her fleet mates (she sacrificed speed for size and luxury), *Aquitania* was by no means slow, and could make the crossing fast enough to allow her to be paired with the smaller, faster *Lusitania* and *Mauretania*.

During the hostilities of the First World War, most passenger liners took leave from their peacetime services and were called into military service. The Cunarders

RMS *Mauretania was the fastest ocean liner of her day. (Bill Miller Collection)*

were no exception, with *Mauretania* and *Aquitania* being requisitioned as both troop and hospital ships. *Lusitania* remained in passenger service, but was torpedoed and sunk by a German U-boat on 7 May 1915.

After the war, Cunard found themselves one ship short for their primary transatlantic service. The loss of the *Lusitania* meant that their weekly service could not be achieved. Help arrived in the way of war reparations and the company was awarded the former Hamburg-Amerika Line flagship *Imperator*, which they renamed *Berengaria*.

These three ships, *Mauretania*, *Aquitania* and *Berengaria*, took Cunard into what is considered their first golden age. With the war behind them, the company now owned the fastest ship in the world (*Mauretania*) as well as the most beautiful (*Aquitania*) and one of the largest (*Berengaria*). Nothing, or so it seemed, could beat Cunard.

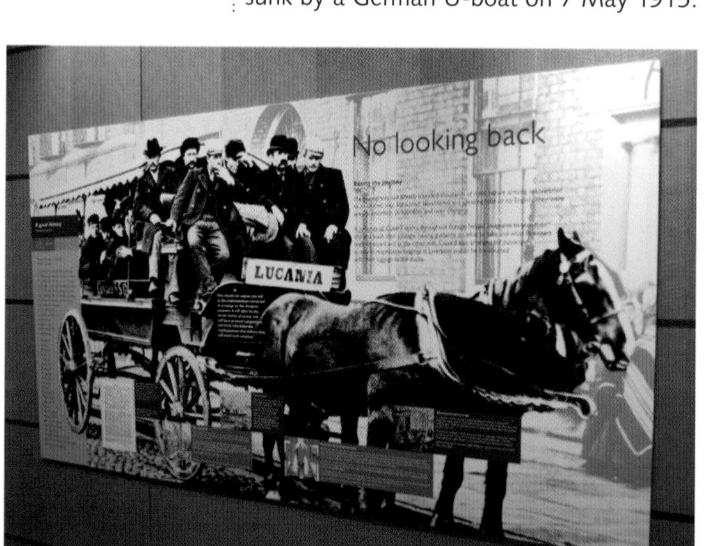

Cunard's early history is illustrated aboard QM2 in the 'Maritime Quest'. (Authors' Collection)

Cunard enjoyed a decade of good times with their three premier liners, however by the 1930s the veteran fleet was beginning to show its age. Advances in naval architecture resulted in a resurgence of competition on the North Atlantic, with liners the likes of Germany's *Bremen*, Italy's *Rex* and France's *Normandie* easily surpassing the Cunard trio in speed, size and luxury.

Cunard's answer was a ship known as 'Hull Number 534'. This ship was under construction at the famed John Brown shipyard in Clydebank. She was big and was intended to be extremely fast, offering Britain a chance at recapturing the Blue Riband and restoring national pride.

However, the Great Depression had taken a significant toll on British industry and it was not until the Government forced a merger between Cunard and their once rival White Star Line that the newly formed Cunard-White Star had the capital to complete construction.

On 26 September 1934, over five years after she was ordered, the new ship was named *Queen Mary* by her namesake, HM Queen Mary, Queen Consort of King George V.

After further months of fit out and trials, *Queen Mary* entered service with Cunard-

Did you know?

QM2 is the longest, largest, most expensive ocean liner ever built.

RMS Queen Mary *was a much loved Cunarder. (Colin Hargreaves)*

White Star as the flagship and after several crossings she won the Blue Riband from the French Line's *Normandie*.

Queen Mary was joined by a sister ship, the *Queen Elizabeth*, which was intended to allow Cunard to meet their long held ambition of a two-ship weekly transatlantic service. However, this goal was stalled with the start of the Second World War, which saw both of the Queens requisitioned for extensive trooping duties.

After the war, Cunard finally realised their dream and placed both Queens on the North Atlantic run. *Mary* and *Elizabeth* captured the hearts and minds of the travelling public and became icons of international travel. During the 1950s Cunard's Queens were considered the only way to travel and hosted a range of celebrities, dignitaries and royalty.

It was with the introduction of the passenger jet that Cunard, along with every other shipping company, faced a new, unbeatable threat. Transatlantic passenger shipping collapsed when the public unanimously moved to the faster, more efficient mode of travel.

Cunard was faced with a significant problem. Their ageing Queens were losing money. The company dabbled in air

◄ *The original* Queen Mary *departing Southampton in the 1960s. (Colin Hargreaves)*

▼ *Until* QM2, Queen Elizabeth *was the largest Cunard ship ever built. (Bill Miller Collection)*

CUNARD

transport forming joint ventures with both BOAC and Eagle Airways. However, the company was unwilling to admit defeat, so in the offices of the iconic Cunard Building in Liverpool plans were formed for the replacement of the Queens.

These plans eventually resulted in the *Queen Elizabeth 2*. Sporting a dual-purpose design which allowed her to perform both the transatlantic service as well as take up winter cruising, *QE2* became known as the 'last of the great transatlantic liners'.

The magnificent QE2 *was the Cunard flagship from 1969 to 2004. (Authors' Collection)*

Cunard's name was carried by QE2 *for forty years. (Authors' Collection)*

Despite being the sole survivor of a bygone era, *QE2* became a worldwide celebrity. Her graceful design and elegant onboard ambiance offered passengers a doorway to the past – a lifestyle seldom found elsewhere in the age of the jet.

Cunard's gamble paid off, in fact at the time of her retirement in 2008 *QE2* had become the longest-serving ocean liner ever.

Shortly after *QE2* entered service, Cunard was purchased by the Trafalgar House Company. This meant that for the first time in the company's long history, they were not independently owned.

For the next three decades the Cunard fleet underwent a number of significant changes. Rather than building new ships, the company opted to acquire existing tonnage which included the highly rated *Sagafjord* and *Vistafjord*, as well as the Sea Goddess Line and along with it, their ships *Sea Goddess I* and *Sea Goddess II*.

By the mid-1990s *QE2* was approaching her thirtieth birthday. Cunard Line's owner, the Trafalgar House Company, was in financial difficulty. Trafalgar House was sold to Kværner, who had no real interest

➤ *By the mid-1990s,* QE2 *was approaching her thirtieth birthday. (Cunard Line)*

➤➤ Vistafjord *was a favourite among regular Cunard passengers. (Bill Miller Collection)*

VISTAFJORD

CUNARD

Of the many ships being built, this Cunard ocean liner is perhaps the most intriguing. She will be the most famous ship in the world even before she takes to the seas. The level of interest is beyond our wildest imagination.

Larry Pimentel, Chairman and COO of Cunard Line, discussing *Queen Mary 2*.

in the shipping line, and many thought that the age of the ocean liner would soon come to an end.

Fortunately this was not to be the case. In April of 1998 Carnival Corporation announced its intention to purchase the Cunard Line. Once the acquisition was complete, Carnival set about reorganising the Cunard brand. *QE2* was partnered with fleet mate *Vistafjord* (which was renamed *Caronia*) to form the basis of the new Cunard Line, while the other ships (at that time, *Sea Goddess I*, *Sea Goddess II* and *Royal Viking Sun*) were transferred to Seabourn Cruise Line.

The intention was to reorganise the company to allow twenty-first-century

▲ *Unlike* QM2, QE2 *is small enough to transit the Panama Canal, as seen here. (Cunard Line)*

◄ QE2 *and* Caronia *together off Barbados. (Cunard Line)*

Did you know?
QM2 carries the designation RMS (Royal Mail Ship) and often carries a tribute bag of mail.

travellers to experience the golden age of ocean liner travel and the existing fleet was refurbished to emphasise the onboard atmosphere of stately British ocean liners.

One week after the purchase was finalised, the industry was stunned by the announcement that Cunard would build a new ocean liner. Code-named Project *Queen Mary*, this was the first true ocean liner to be built since *QE2*'s introduction to service back in 1969.

In fact, the reasons for Carnival's purchase of Cunard Line were closely interconnected with Carnival chairman, Micky Arison's, desire to build the first ocean liner of the twenty-first century.

The announcement of Project Queen Mary *gave reassurance for the future of the Cunard Line. (Authors' Collection)*

With the announcement of Project *Queen Mary*, Cunard formed a design team to work on creating the first Atlantic liner since *QE2*. Stephen Payne was appointed as the head architect for the very ambitious project, having long been a lover of ocean liners, with the great Cunard Queens being amongst his favourites.

In May 1998 Payne and colleague Richard Moore began a two-year study to determine the form that the new ship would take. One of Payne's first acts was to sail aboard *QE2* to experience a transatlantic crossing. Commodore Warwick recalls the time spent on *QE2* and how it influenced *QM2*'s overall design:

> Shortly after it was announced that Cunard Line was proposing to build the *Queen Mary 2*, the architect, Stephen Payne sailed with me on the *Queen Elizabeth 2* to experience a transatlantic crossing on a liner. This crossing was the start of a dialogue between the designers and officers of the ship to consider the

The cruise industry has changed significantly since the days of the transatlantic liner. (Authors' Collection)

➤ *Steel cutting takes place for the new QM2. (Pam Massey)*

QM2 *towers over the dry dock during construction. (Pam Massey)*

various aspects of a liner versus a cruise ship. Over the ensuing months many of the characteristics of a liner were discussed. These included features that could be influenced by the weather and speed, such as the need to have a longer

➤ *QM2's long bow showing her classic lines. (Pam Massey)*

foredeck than a cruise ship, the size of the wheelhouse windows and the height of lifeboat stowage above sea level.

When the first draft outline plans were received these led to further more detailed discussion on matters like gangway and launch embarkation points, capacities of fresh water and bunkers, ballasting arrangements and Bridge equipment. During one of these discussions it was proposed to add the forward suite rooms which were not on the original plan.

The ship had to be strong, far stronger than a cruise ship, while also offering the amenities that passengers expected from modern cruise vessels. Strength aside, there were other significant considerations that needed to be incorporated into the ship. She had to be fast enough to operate the

◀ *QM2 in the yard on a rainy day. (Pam Massey)*

Did you know?

QM2 is the Cunard Line flagship.

The Bridge has been installed giving QM2 her distinctive appearance. (Pam Massey)

transatlantic crossing in six days, while having the long bow of an ocean liner, with superstructure set back. The ship also had to incorporate lifeboats higher in the superstructure than on a contemporary cruise ship (to protect them from the ferocity of the North Atlantic), while the hull would require a deeper draft to ensure stability.

QE2's superstructure had been built using aluminium. This decision had allowed excellent stability, while lightening the ship

◄ The stabilisers are installed. (Pam Massey)

◄◄ QM2's stern towers over the dry dock. (Pam Massey)

Cunard lettering is now visible. (Pam Massey)

and thus reducing the draft. Furthermore, the designers of *QE2* had been able to incorporate an extra deck into the ship due to the aluminium's weight saving aspects. However, while *QE2* was an immensely strong ship, over her forty-year service, the aluminium superstructure had required increasing levels of care to ensure it was able to withstand the forces applied to it during annual Atlantic service.

Given that *QM2* was intended to sail for forty years, the decision was made that she would be built with a steel superstructure. This, coupled with the demands for a greater number of balcony cabins (which the travelling public expected), resulted in the ship's design ballooning to become the largest passenger ship in the world, at the time of her introduction into service. The final dimensions were 1,132ft (345m) by 135ft (40m), 169ft longer than *QE2* and 1.28 times the length of *Titanic*!

Significant consideration had to be taken when selecting a shipyard to build the new ship and a number of shipyards were considered. Shipbuilding techniques had changed dramatically since *QE2*'s construction back in the late 1960s and so too had the shipbuilding industry. The previous three Queens had all been constructed in Scotland, at the John Brown (later Upper Clyde) shipbuilders in Clydebank, which had long since closed down.

The only British shipyard that was able to tender for the work was the historic Harland and Wolff Company in Belfast, Northern Ireland. Harland and Wolff are widely remembered as the builders of the *Titanic*, but they were also responsible for

Did you know?

Once around *QM2*'s 360-degree Promenade Deck is 2,034ft, which is well over a third of a mile.

QM2's sheer size is visible in this shot. Her mast is on the dockside. (Pam Massey)

the construction of the Orient Line's *Oriana* and P&O's *Canberra*.

Other yards throughout Europe were also considered, with a notable contender being Chantiers de l'Atlantique. This French shipyard had an enviable pedigree that included the immortal *Normandie* and also the *France*, which was later converted to become Norwegian Cruise Line's *Norway* (the world's longest passenger ship until *QM2*'s introduction into service). They were also highly active in modern shipbuilding techniques, having constructed a number of cruise ships, including Royal Caribbean's *Sovereign of the Seas* which was at the time of her construction the world's largest cruise ship by tonnage.

The combination of ocean liner experience, coupled with their current strength in modern passenger liners, gave Chantiers de l'Atlantique the edge over the financially troubled Harland and Wolff. On 10 March 2000 a letter of intent was signed with the French shipyard, with the formal contract being signed by Micky Arison (Chairman and CEO of Carnival Corporation) and Patrick Boissier (President ALSTOM Marine and Chairman and CEO of Chantiers de l'Atlantique) on 6 November.

The ship was thereon known as G32, the yard number allocated to the future Queen. At the same time as the contract was signed, Cunard unveiled the interior design for *QM2* in London, showcasing the intended interiors of the world's largest liner. '*QM2* will be strikingly different from modern hotels on the water,' commented a Cunard spokesperson on the announcement of the shipbuilder.

Did you know?

QM2 is the only ocean liner currently offering scheduled transatlantic crossings.

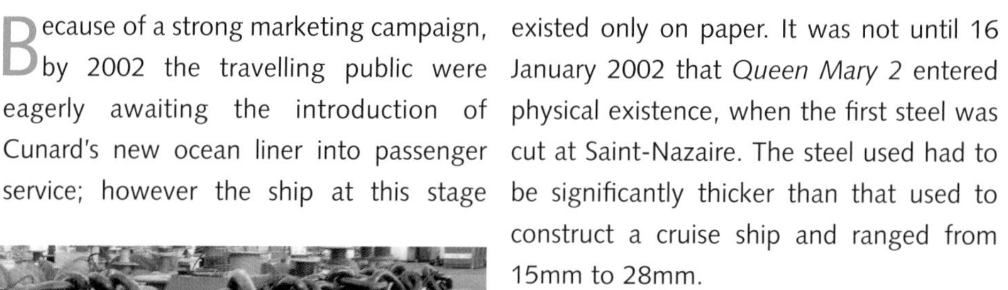

Because of a strong marketing campaign, by 2002 the travelling public were eagerly awaiting the introduction of Cunard's new ocean liner into passenger service; however the ship at this stage existed only on paper. It was not until 16 January 2002 that *Queen Mary 2* entered physical existence, when the first steel was cut at Saint-Nazaire. The steel used had to be significantly thicker than that used to construct a cruise ship and ranged from 15mm to 28mm.

QM2's anchor chains awaiting installation. (Warwick Family Collection)

Pamela Conover, President of the Cunard Line, gave a speech to the gathered press, expressing the significance of *QM2* for twenty-first-century cruising, before pressing the button that begun the steel cutting process in earnest:

> I am absolutely delighted to officially initiate the construction of the successor to such great transatlantic liners as *Queen Mary*, *Queen Elizabeth* and *QE2*. This begins a new era in the history of Cunard.

QM2's mechanical areas during construction. (Warwick Family Collection)

Marketing of QM2's first voyages started well before the ship was complete. (Pam Massey)

The following months saw continuous cutting of the steel plates that would form *QM2*. Unlike ocean liners of days gone by, which were built from the bottom up and riveted or welded together plate by plate, *QM2* was constructed by connecting a number of large prefabricated blocks together. This technique has become the mainstay of modern passenger ship construction, where the finalised design is divided up into blocks and like a giant Lego

Captain Connection

Many people who have sailed aboard *QM2* know the name Commodore Warwick, but did you know there have been two Commodore Warwick's in Cunard's history?

Commodore William E. Warwick had been a relief captain to *Queen Mary* and *Queen Elizabeth* and was selected to standby *QE2* during her construction. In 1969 he became that ship's first captain.

His son, Commodore Ronald W. Warwick also commanded *QE2*, when in 1990 he relieved Captain Woodall in anticipation of a visit aboard by HM Queen Elizabeth II. This was the first time that a father and son had commanded the same vessel in Cunard's long history.

Commodore Ronald W. Warwick was later appointed to stand by *QM2* and became her first master, meaning that both *QM2* and her predecessor *QE2* set off on their maiden voyages with a Warwick in command.

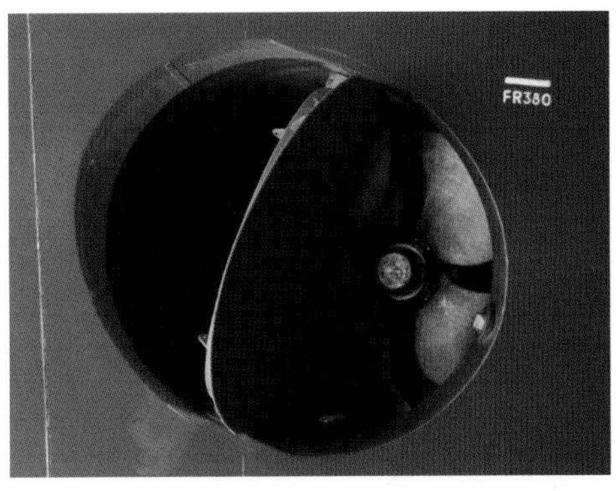

◄◄ QM2 *has four stabilisers, making for a pleasant journey. (Emily Wealleans)*

◄ *To keep her hull streamlined,* QM2 *has doors over her bow thrusters that open when in use. (Emily Wealleans)*

set, is constructed piece by piece to form the finished ship. *QM2*'s keel formed the first of these blocks and on 4 July 2002 the ship's keel was ready to be laid.

Pamela Conover invited Captain (later Commodore) Ronald W. Warwick to perform the honour of signalling to the crane driver permission to lower the ship's keel into place. Via a walkie-talkie he said, 'Crane Driver, this is Captain Warwick speaking. Please commence the building of my ship', after which the 650 tonnes of steel (forming sections 502, 503 and 102) were lowered into place.

Thus, *Queen Mary 2* began to grow, becoming a towering landmark to the town of Saint-Nazaire.

Given the complexities of building a ship the likes of *QM2*, it was a relatively short time before she was ready to be floated

➤ *A view forward showing one of QM2's two bow anchors. (Emily Wealleans)*

➤➤ QM2's *impressive breakwater was a design trait of the* Normandie. *(Authors' Collection)*

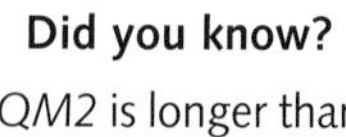

Did you know?

QM2 is longer than three football fields and half as long again as the Canary Wharf Tower is high.

Looking forward from atop QM2 at the town of Saint Nazaire. (Warwick Family Collection)

Did you know?

QM2 took over as flagship from *QE2* in May 2004.

out. Floating a ship for the first time is the modern-day equivalent of a launch. Water is pumped into the dry dock where the ship has been built which eliminates many of the uncontrollable factors associated with sending a ship down the slipway to meet the sea.

QM2's hull met its natural element for the first time on 1 December 2002. By this stage, the diesel engines had been installed deep within the hull, lowering the ship's centre of gravity and increasing stability. Once afloat the ship was moved to a fitting out dock in order for the work to continue.

While in the fitting out dock *QM2* underwent a radical transformation. Although outwardly the ship had achieved much of her awe inspiring size, it was the transition of her interior, from an empty hulk to a floating palace, that resulted in the magnificent ship we see today.

Cunard wanted their new monarch to be simple for passengers to navigate and as such designed her around four major stairwells, all of which spanned from top-to-bottom.

The ship's interior had to include a number of signature public rooms which had become extremely popular aboard previous Cunarders, such as *QE2* and *Caronia*. A ballroom, a traditional English pub and alfresco dining options were essential, as well as sufficient bars and lounges to accommodate the passengers.

Drawing from the format of *QE2*, where restaurants and cabin categories are connected, *QM2*'s design incorporated three formal restaurants. The Queens Grill and Princess Grill, located at the aft

The Commodore Club nears completion. (Warwick Family Collection)

A look inside the Pavilion Pool during construction. (Warwick Family Collection)

end of deck seven cater for those paying the highest tariff, while the majority of passengers dine in the Britannia Restaurant.

The size of *QM2* allowed for extravagant interior spaces to be created. *QM2*'s 135ft width allowed for the largest dance floor at sea in the Queens Room. Her 1,132ft length allowed for Illuminations, the first planetarium at sea, in addition to a dedicated show lounge called the Royal Court Theatre.

◄◄◄ *The completed Britannia Restaurant. (Authors' Collection)*

◄◄ *The completed Commodore Club. (Authors' Collection)*

◄ *The Queens Room, QM2's ballroom. (Authors' Collection)*

Queen Mary 2

The interior fit-out took hundreds of man hours to complete and many different trades were employed to transform the ship. Specialist trades and artists were also employed. This included model maker Henk Brandwijk, who created the stunning model of *QM2* that resides in the Commodore Club and John McKenna, whose aluminium bas-relief welcomes guests upon entering the Grand Lobby.

By September 2003, with the interior works nearing completion, the ship was ready to go to sea. Her sea trials began on 25 September, when for the first time the world's largest ocean liner set sail under her own power.

Aboard *QM2* a mix of Cunard and shipyard representatives were present. The shipyard sent a 400-man crew (under the command of a French captain) aboard, while Captain Warwick, Stephen Payne and a number of other Cunard personnel were aboard to observe the trials.

QM2's first real voyage was a notable event for the town of Saint-Nazaire and large crowds of proud townspeople gathered to view the ship's departure.

The ship was at sea for three days, and testing was completed on her mechanics. Speed, crash turns, full astern manoeuvres, anchor dropping and retrieval as well as stabilisers were all extensively tried and tested. The first trials were a triumphant success, with *QM2* achieving all of her set targets, much to the delight of Cunard!

The Grand Lobby. (Authors' Collection)

QM2 *towers above the dry dock. (Warwick Family Collection)*

QM2 *has three bow thrusters, giving her excellent manoeuvrability. (Warwick Family Collection)*

◀ *QM2's podded propulsion system is state of the art. (Warwick Family Collection)*

Did you know?

QM2 is five times longer than Cunard's first ship, *Britannia*.

Pamela Conover was President and COO of Cunard Line during QM2's construction:

I feel very lucky and privileged to have been a part of this project from inception through the launch. It was Cunard's position of operating transatlantic liners which first attracted Carnival Corp. The essence of the brand was built around that history however it was clear that a new liner would have to be built given the age of the *QE2*.

Immediately following the acquisition by Carnival of Cunard, our Chairman Micky Arison started discussions with the architects and shipyards about what it would take to build a modern transatlantic liner. So many people played key roles in the concept and design and then in the construction itself. I just feel incredibly lucky to have been a part of the process from the design, to my cutting the first piece of steel, to hosting Her Majesty the Queen to christen her.

As a British person I am very proud of the first British flag transatlantic liner to have been built in thirty-five years. It is also very gratifying to see that she has been such a commercial success and it is wonderful to still hear from people who are sailing on her for the first time what a great experience they had.

Did you know?
Commodore R.W. Warwick was *QM2*'s first captain.

The owner's trials took place in November 2003. During these trials, *QM2*'s official speed was calculated at 29.62 knots, making her the second fastest merchant vessel afloat (after *QE2*'s 32.5 knots). Encouraging for Cunard, however, was that *QM2*'s speed could easily allow her to complete six-night crossings.

On 15 November tragedy struck Saint-Nazaire when a gangway attached to *QM2* collapsed, killing fifteen and injuring twenty-five. Those on the gangway formed part of a group of special guests, shipyard workers and their families, invited to tour the ship prior to her handover to Cunard. The town was in mourning and local hospitals were near capacity attending to the injured.

Despite this tragic setback, the remaining workers aboard *QM2* completed their duties with the utmost professionalism; such was the pride that the people of France had for the mammoth ship they had created.

Ronald Warwick stands in front of his new command. (Warwick Family Collection)

The completed QM2 at Southampton prior to her naming ceremony. (Pam Massey)

Looking marvellous, QM2 prepares for her maiden voyage. (Pam Massey)

➤ *Commodore R.W. Warwick with HM Queen Elizabeth II. (Commodore R.W. Warwick)*

A significant day for Cunard was 22 December; this day saw them take delivery of *QM2*. A brief ceremony took place, whereby the French flag was lowered from the ship's mast. Captain Warwick's command of 'Hoist the Red Ensign' signified *QM2* becoming a Cunarder and the largest ship ever to grace the British merchant navy, as both the Red Ensign and the Cunard house flag took pride of place. *QM2* departed France the next day, bound for Britain, wearing a large banner that read 'Thank you Saint-Nazaire' as a gesture to the town that had created her.

She arrived to a jubilant welcome in Southampton on Boxing Day, 2003. While alongside, hoards of journalists, travel agents and other invited guests inspected the ship.

Her Majesty, Queen Elizabeth II officially named *Queen Mary 2* on 8 January 2004. Prior to the ceremony the queen toured the ship and signed the visitor's book in the library. A ceremony that included a video presentation about the Cunard Queens and a blessing from the Right Reverend Michael Scott-Joynt, the Bishop of Winchester, preceded the official naming which took place before Prince Phillip, Micky Arison, Pamela Conover and the newly designated Commodore Warwick, as well as many dignitaries and invited guests.

By daybreak on 9 January, *QM2* was officially in service with Cunard. She remained in Southampton for a further three days, as final preparations and storing took place. Finally, the first passengers stepped aboard before the ship set sail on her maiden voyage to Fort Lauderdale.

Did you know?

QM2 is three times as long as St Paul's Cathedral is high.

◄◄ *QM2 will carry the Cunard name well into the twenty-first century. (Authors' Collection)*

◄ *After years of planning, by 2004* QM2 *was finally a Cunard ship. (Authors' Collection)*

Despite being the largest ocean liner ever built, as well as the most significant new build in Cunard's recent history, *QM2* was not named flagship of the Cunard Line at the time of her introduction into service. This honour remained with *QE2* for several more months, while *QM2* 'earned her stripes'.

During the first four months of Cunard service, *QM2* undertook a number of

➤ *The new QM2 enters New York. (Russ Willoughby Collection)*

◄ *QM2 with the iconic New York tugs. (Russ Willoughby Collection)*

Did you know?

QM2 covers an area of 3.5 acres.

pleasure cruises. Her maiden voyage had been a magnificent success, despite the lack of tradition in the route chosen (all other Queens' maiden voyages had been a direct run to New York), with the voyage taking her to Fort Lauderdale via Funchal, Tenerife, Las Palmas, Bridgetown and Charlotte Amalie.

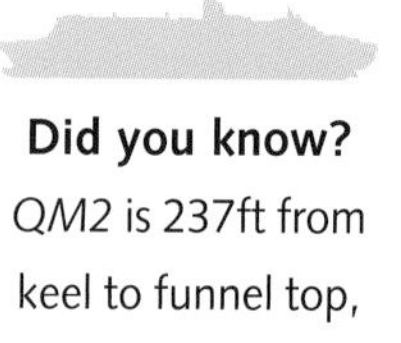

Did you know?
QM2 is 237ft from keel to funnel top, towering over 200ft above her waterline.

➤ QM2 *berths in New York City. (Russ Willoughby Collection)*

Her early voyages were fully booked, thanks partly to excellent PR and marketing, but also due to *QM2*'s natural ability to attract interest from the general public. Voyages included calls at Rio de Janeiro (for Carnaval), a number of Caribbean cruises, and a long voyage back to Southampton via Western Europe. The ship also completed

QM2 *with fleet mate* QE2 *in New York. (Thad Constantine)*

◀◀ *Two Queens together in New York. (Sam Warwick)*

◀ QM2 *as seen from* QE2 *during the tandem crossing. (Kyle Johnstone)*

The QM2 at sea during the tandem crossing with QE2. (Kyle Johnstone)

her maiden (direct) transatlantic crossing which departed Southampton on 16 April, arriving in New York on 22 April.

On 25 April *QM2* was met by her fleet mate, *QE2*, and for the day the two liners sat side by side at the West Side piers in Manhattan. That evening the two iconic Cunarders slipped their moorings and set sail on a tandem Atlantic crossing, keeping each other in sight throughout the voyage, to the delight of the passengers.

Upon arrival in British waters, the ships were met by an RAF Nimrod jet that undertook a flyover. They then proceeded into Southampton waters and took rest at their respective berths.

The 2.5ft-tall Boston Cup (originally called the Britannia Cup) was created in Boston for presentation to Sir Samuel Cunard as a thank you gesture when Cunard selected Boston as the American terminus for the transatlantic mail service in 1840.

The whereabouts of the Boston Cup is unknown for much of its history, but in 1967 it was rediscovered in a Maryland antique shop and purchased by Cunard Line. In 1969 it was placed aboard *QE2* and remained there until 2004.

On 1 May 2004, as the two Cunarders sat in Southampton, a ceremony took place aboard *QE2* where Captain Ian McNaught of *QE2* handed the cup over to Commodore Ronald Warwick of *QM2*, signifying the transfer of flagship status to *QM2*.

The newly crowned flagship enjoyed a spectacular first year of service, highlighted by a number of significant maiden arrivals. One such voyage took her to South Queensferry (for Edinburgh), Geiranger

Did you know?

QM2 is three and a half times as long as the tower of Big Ben is high.

The two Queens together in Fort Lauderdale, Florida. (Kyle Johnstone)

Fjord, Alesund and Bergen, where she attracted considerable attention.

QM2 then sailed to Hamburg in Germany, where she was met by her largest crowd yet. Well over 1 million people lined the banks of the River Elbe to see her. As the call was an overnight stay,

▲ *The Boston Cup is handed over aboard* QE2, *signifying the start of* QM2*'s reign as flagship. (Cunard Line)*

◄ *Two icons of engineering in close quarters. (Authors' Collection)*

▲ *QM2's maiden call in the breathtaking Geirangerfjord, Norway, was shared with* Marco Polo. *(Authors' Collection)*

spectators camped out near the terminal to reserve the best vantage points from which to watch the sail away the next morning.

Her final call on this voyage was to Rotterdam, where again she was met by a large crowd. She returned to Southampton and disembarked her passengers, all of whom went home knowing they had been part of history.

In 2004 the games of the XXVIII Olympiad took place. Held in Athens, the games were a significant moment in *QM2*'s history, as the ship was used as a floating hotel for the duration of the games.

Peter Shanks, Senior Vice President Cunard Line (Europe), said: 'Cunard Line is delighted to be a part of this historic event in Greece, the birthplace of the Olympic Games. Being selected by the Athens organising committee is a great honour and the first step in a very exciting process.'

QM2 was docked in Piraeus and played host to former US President George H. Bush, then British Prime Minister Tony Blair, then French Prime Minister Jacques Chirac, as well as the US men's basketball team.

An unusual honour was bestowed upon *QM2* during 2005 when she was selected to carry the first US signed copy of J.K. Rowling's book *Harry Potter and the Half Blood Prince*. The book was transported in a steamer trunk via armoured guard from the author's Edinburgh office before being loaded aboard *QM2* for the six-night transatlantic. Once safely in New York, the book was transported to the gala book launch event.

A well-known Cunard personality joined *QM2* as Entertainment Director in 2005. Alastair Greener had previously served aboard *QE2* and has since worked on *Queen Victoria* and the new *Queen Elizabeth*. Of his appointment to the Cunard flagship he said:

Geirangerfjord is a sight that none of the passengers aboard QM2 *will forget. (Authors' Collection)*

Did you know?

QM2 is over two and a half times longer than the height of the 'London Eye'.

I'm sure everyone who has ever sailed on *Queen Mary 2* will remember their first view of this truly impressive liner. Unique is an over used word these days, but is one that absolutely fits *Queen Mary 2*.

I have been on over twenty five ships during my career at sea, but walking into *Queen Mary 2*'s Grand Lobby in 2005, I still recall that initial impression of rich elegance and understated grandeur as I

QM2 *alongside in Hamburg with an impressive number of spectators. (Authors' Collection)*

Thousands gather in Rotterdam to view QM2*'s departure. (Authors' Collection)*

walked down that stunning staircase on the thick deep red carpet. I thought to myself, 'Now I understand why everyone says this ship is so special'.

QM2 berthed in Rotterdam. (Ben van Zeijl)

An eventful year for *QM2* was 2006, which started with an extended line voyage from Fort Lauderdale bound for the US West Coast. As the ship is too large to transit the Panama Canal, this voyage included the transit of Cape Horn.

Unfortunately, during her departure, the ship struck a channel wall which damaged one of the high-tech propeller pods. *QM2* was able to sail (at a reduced speed) on her remaining three pods, however the length of the journey meant that the call at Rio de Janeiro had to be missed in order to keep on schedule. A number of passengers aboard, disappointed at missing the call at Rio, threatened a sit-in, resulting in Cunard offering compensation.

Having transited Cape Horn in early February, *QM2* made a notable call at Long Beach on 23 February. There she met her namesake RMS *Queen Mary*, the elder Cunarder having resided in the Californian city since her retirement in 1967. The two *Marys* exchanged lengthy whistle blows to mark the occasion.

Due to her reduced speed, Cunard made a number of alterations to *QM2*'s schedule until she was able to visit the Blohm & Voss shipyard in Hamburg, Germany, for repairs. The work was completed in two stages, with the damaged pod first being removed in June 2006 before being returned to the ship in November that same year.

The latter dry-docking time was also used to install sprinkler systems to all of the ship's balconies – a new safety requirement introduced after fire broke out aboard the *Star Princess*. Another subtle yet important alteration made to *QM2* during her November dry-docking was the extension of the ship's Bridge wings in order to improve visibility. Each wing was extended by 2m, resulting in an uninterrupted span of 44.95m from port to starboard.

▲ *Boats gather for QM2's send-off in Rotterdam. (Ben van Zeijl)*

QM2's propeller pod. (Emily Wealleans)

The sheer size of QM2 is overwhelming. (Emily Wealleans)

Christel Hansen is the captain's secretary aboard QM2:

My work is continually challenging and interesting and when I add to that the fabulous camaraderie of the second family I work with, I feel privileged to be on the magnificent *Queen Mary 2*.

With the popularity of *QM2*'s longer voyages in 2005 and 2006, the decision was made to send the world's largest ocean liner on a world cruise. Dubbed 'Around the World in 80 Days', *QM2*'s maiden circumnavigation of the globe was undertaken while fleet mate *QE2* was completing her 'Silver Jubilee World Cruise'. Cunard made the most of these voyages, which saw the two great liners meet in a number of ports around the world.

The first such meeting took place in Fort Lauderdale before the two liners parted ways, *QE2* headed for the Panama Canal while *QM2*, too large to transit the canal, made her way south, bound for Cape Horn.

Upon rounding the Horn, *QM2* sailed north for an iconic maiden call at San Francisco. Arriving on 4 February amid spectacular fanfare, the ship remained overnight and set sail the following evening bound for the southern hemisphere.

QM2's stern towers over the Honolulu dockside. (Authors' Collection)

Safely docked, QM2 basks in the tropical heat of Pago Pago. (Authors' Collection)

The world's largest ocean liner was the tallest man-made structure in Pago Pago. (Authors' Collection)

➤ *A view of Auckland, New Zealand, from QM2's deck. (Authors' Collection)*

Enroute 'down under', the ship visited Hawaii for the second time, before calling at Pago Pago, American Samoa. Here *QM2* was by far the tallest ship to attempt docking in the island port.

Captain Christopher Rynd, who commanded *QM2* for this segment of the world cruise, recalls docking *QM2* in Pago Pago for the first time:

> My concern for the first arrival in the port of Pago Pago in American Samoa was whether she would fit under the cable car wire that spans the harbour. Pago is a perfect natural harbour surrounded by steeply rising land with a cable car running from the town across the harbour to a high peak were the TV broadcast aerial is. No one had information on the clearance under this wire and this was the tallest (highest air draft) vessel ever to visit. In the end the First Engineer was positioned on the top of the funnel, the highest point at 62 metres, and we approached slowly while he assessed if the ship would clear. It did, narrowly.

Auckland, New Zealand, saw another iconic maiden arrival while, on 20 February 2007, *QM2* was involved in a long-awaited event in Sydney Harbour. Here, for the first time since the Second World War, the largest Australian city would host two Cunard Queens.

Noted maritime historian and author Bill Miller recalls this very special event:

> The Queens always seem to be making news. But in February 2007, on a rather magical Australian summer's evening in

vero

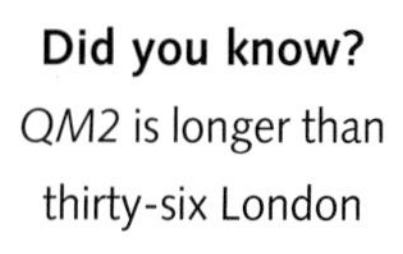

Did you know?
QM2 is longer than thirty-six London buses.

Sydney, Cunard could not have even wished for a more successful occasion. The two Queens – the *QE2* and the *QM2* – met while travelling westward on their respective world cruises. We had arrived that morning (from San Francisco) to a joyous, spirited reception aboard the 1,132ft-long *Mary 2*.

While a Tuesday and so a normal work day, the crowds along the shores were huge. Hundreds of boats serenaded the liner from daybreak onwards, just after we passed through the famed Sydney Heads. Helicopters and small planes buzzed overhead, fireboats sprayed and all while whistles, horns and sirens created a seemingly endless concert. Onboard, the outer, upper decks were crowded with passengers and crew alike. Slowly, the 151,000-ton Queen made her way into the harbor, passed that grand Opera House and then, with easy precision, turned just before the Harbour Bridge, reversed course and then slowly docked in a specially prepared berth at the naval dock yard on Garden Island. She was simply too big, too long, to use the normal passenger berth at Circular Quay, located just across from the Opera House. Once ashore, the 2,500 or so passengers (including 800 Australians) were bussed across downtown Sydney to a cruise terminal in Darling Harbour for final disembarkation.

Sydney newspapers and TV news were ablaze with the *QM2*'s arrival. She was not just headlines, but big headlines. Some Australians remembered, of course, that wartime meeting of the original *Mary* and *Elizabeth*, then both in troopship gray, off those Sydney Heads

back in January 1941. Afterward, for a year so, the converted Queens carried Aussie soldiers off to war, mostly across the Indian Ocean to Suez for the Allied reinforcement of North Africa and the Middle East.

QM2 *passes the Sydney Opera House. (Richard Edwards)*

▲ *Many people took boats out to meet* QM2 *in Sydney, Australia. (Richard Edwards)*

➤ QM2 *is met by a Port of Sydney fire-boat. (Authors' Collection)*

Throughout the day, there were luncheons, tours, special tea parties and a long string of televised tours and interviews with top staff members onboard the *QM2*. But an even bigger, greater occasion was just hours away. At seven that night, as Sydney skies were slowly, evocatively turning from golden hues to cinnamon and finally to charcoal, we boarded a friend's large cabin cruiser and joined an estimated 500 other craft in the outer harbour. With the towering *QM2* still berthed at the naval dock yard (and sailing that night close to midnight

➤ QM2 *and* Carnival Victory. *(Thad Constantine)*

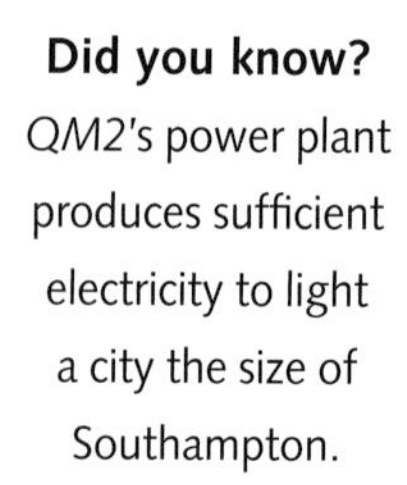

Did you know?

QM2's power plant produces sufficient electricity to light a city the size of Southampton.

for Hong Kong), the majestic *QE2* arrived, flag-bedecked and sounding her throaty, big liner whistles. Again, small craft formed the royal procession. Horns and sirens again sounded continuously and those fireboats sent up their great watery fans. Off the naval pier, the 70,000-ton *Elizabeth* seemed to pause, as if to pay homage, a sort of royal curtsey, Queen Mother to Queen so to speak, to her larger successor.

With their whistles, they exchanged greetings. They had, in fact, been together a month before, in mid January, as they sailed together from Fort Lauderdale. Then, slowly, graciously, now beginning to light from end to end, the *Elizabeth* made her way toward the Harbour Bridge, passing the floodlit Opera House and then, with tugboat assistance, was swung round, the front of her 963ft-long hull pointed outward, and berthed at Circular Quay, the main passenger terminal.

It was all perfectly timed, choreographed as if a part of the royal ballet, and then finished off with a blazing fireworks show (some said 'the greatest ever in Sydney harbour'). Two million were reportedly along the shores or on boats and there were subsequent traffic jams, going well into the wee hours that took hours to sort. But in all, it was a grand occasion, a thoroughly memorable travel experience and a gala meeting of the world's two most famous liners. Cunard must have been thrilled.

Seven days later the ship called at Hong Kong, before making way to Singapore and

CARNIVAL VICTORY
Carnival Victory

Did you know?

QM2 is twice as long as the Washington Monument is high.

an overnight call at Cochin, India. Although *QM2* is too large to transit the Panama Canal, she was able to transit the Suez Canal homeward-bound to Southampton.

The success of this initial world cruise has seen *QM2* undertake further world cruises, resulting in her popularity growing in countries all over the world.

The ship's reputation was further enhanced when the popular National Geographic television show *Megastructures* was filmed aboard the ship. Viewers got a first-hand look at life behind the scenes on Cunard's flagship, including a tour of the recycling centre and a look backstage in the Royal Court Theatre.

➤ *All of QM2's cabins offer double beds and are serviced twice daily. (Authors' Collection)*

The year 2008 was momentous for the Cunard Line. In June the previous year the company shocked the world when they announced the sale of their famed *QE2* to Dubai World, with the date of her departure being November 2008.

After an iconic forty years of service, the farewell season for *QE2* was immense, with a depth of emotion not seen since the retirement of the original *Queen Mary* forty-one years earlier. As the flagship of the Cunard Line, and successor to *QE2*, it was appropriate for *QM2* to be involved in the farewell celebration of her predecessor.

Cunard rearranged schedules to allow for the three Queens (*QE2*, *QM2* and *Queen Victoria*) to meet in New York as well as Southampton, while special tandem eastbound and westbound crossings were organised, with both Atlantic Queens sailing within sight of each other. Emotions were high aboard both ships and most passengers agreed that these were crossings they would never forget.

Two Atlantic greats, QE2 *and* QM2, *bid farewell.* Queen Victoria *is visible in the background. (Cunard Line)*

➤ *The final meeting of Cunard's* QE2 *and* QM2 *in Southampton after their 2008 tandem crossing. (Matt Welsby)*

➤➤ *Fireworks as* QM2 *sits alongside in Greenock, Scotland. (Rob Lightbody)*

With *QE2*'s departure from active service, *QM2* was left as the last active trans-atlantic liner, a role her predecessor had assumed for much of her life. Whether another liner will be built is debatable, but as *QM2* is still a young ship she is well

CUNARD

suited to carry the Atlantic baton for the foreseeable future.

Despite the elder Cunarder's retirement, the two ships met again in Dubai. In fact, during *QM2*'s 2009 world cruise the two ships sat side by side at Port Rashid, an event that was repeated in 2010. On both occasions as *QM2* departed the port, the two liners exchanged lengthy whistle blows – an emotional occasion that brought many of those aboard the departing *Mary* to tears.

In October 2009, *QM2* made headlines when she undertook a round-Britain cruise which included maiden calls at Greenock and Liverpool. On both occasions the ship was given a jubilant welcome with significant crowds and special events organised for those passengers lucky enough to partake in the voyage.

Cunard's connection to Greenock is still strong. (Rob Lightbody)

QM2 *departs Fremantle during her 2010 world cruise. (Authors' Collection)*

Did you know?

QM2 is only 117ft shorter than the Empire State Building.

➤ QM2 *dwarfs accompanying boats as she departs Fremantle. (Authors' Collection)*

➤➤ QM2 *sails into the sunset. (Authors' Collection)*

◄◄ *Spectators gather to get a last glimpse of QM2. (Authors' Collection)*

◄ QM2 *docked at Brooklyn. (Matt Welsby)*

The magnificent QM2 at New York's Brooklyn terminal. (Matt Welsby)

Michael Gallagher, Public Relations Executive for Cunard Line, was aboard for the historic Liverpool arrival:

> I was wondering what sort of welcome *Queen Mary 2* would receive in Liverpool and I have to say I was overwhelmed with the response the ship got. It underlines the affection Liverpudlians still have for all things Cunard and the sight of the biggest ship ever to sail up the Mersey was one to behold!

QM2's 2010 world cruise saw the ship continue making history with a number of maiden arrivals, drawing significant onlookers. Her Australian calls, which included inaugural visits to Adelaide and Fremantle, were so popular that Cunard subsequently announced that the ship

QM2 *berths at the Brooklyn Cruise terminal as she is too large to dock on the West Side piers. (Matt Welsby)*

▲ QM2 *making her inaugural entrance to Durban. (Holger Jurgensen)*

would undertake a special circumnavigation of Australia in 2012.

Like all Cunard ships, *QM2* has a reputation for providing high-quality speakers for passenger enrichment and education and this was again evident when, during the 2010 world cruise, Nobel Peace Prize recipient Desmond Tutu boarded the ship in Mauritius.

Tutu spoke to a full house in Illuminations and later disembarked in Cape Town when the ship made her inaugural call to the South African port.

Tim Wilkin works with the Cunard Insights Programme and has organised for many famous speakers to lecture aboard *QM2*:

> How amazing it's been to be part of the *Queen Mary 2*'s story. It's always a thrill to see *Queen Mary 2* in Southampton port standing like a regal lady waiting to take the new guests on voyages of adventure, back to the days of the grand liners. *Queen Mary 2* is a ship like no

other and I am thrilled to be able to add to the guests' experience on board with the Insights programme.

QM2 proved the value of ongoing ocean services when, in April 2010, she was fully booked (with over 1,000 passengers on the standby list) following the Icelandic volcanic eruption that left a plume of smoke over much of Europe and grounded flights for days.

The ship, which was undertaking her scheduled transatlantic crossings, attracted a high volume of booking requests as many stranded air travellers turned to *QM2* as their only option for getting home.

The strength of *QM2*'s appeal grows as each completed voyage leaves new guests in awe of this grand liner. She is a testament to the vision of her builders, of twenty-first-century shipbuilding, as well as a fitting tribute to all those great Cunarders that came before her.

If Samuel Cunard knew of the *QM2*'s story, he would be very proud indeed.

Commodore Warner reflects on being appointed master of *QM2*:

I felt both fortunate and privileged to be appointed Master of *Queen Mary 2* in the summer of 2005. Having spent most of my career with P&O, it had never really crossed my mind that I might have the opportunity to one day command the most famous and greatest ocean liner in the world.

Did you know?

QM2 is 147ft longer than the Eiffel Tower is high.

QM2 SPECIFICATIONS

QM2 GENERAL INFORMATION

Name:	*Queen Mary 2* (*QM2*)
Gross Registered Tonnage:	151,400 tons
Passenger Decks:	14 (13 + deck 3L)
Length:	1,132ft (345m)
Width:	135ft (40m)
Width at Bridge Wings:	147.5ft (45m)
Draft:	32ft 10in (10m)
Height (Keel to Funnel):	237ft (72m)
Builders:	Chantiers de l'Atlantique, France
Keel Laid:	4 July 2002
Floating Out Date:	21 March 2003
Maiden Voyage:	12 January 2004
Maximum Passenger Capacity:	3,056
Standard Crew Capacity:	1,253
Port of Registry:	Southampton, England
Official Number & Signal Letters:	9241061 & GBQM
Owners:	Cunard Line
Ship Building Cost:	Approx. $800 million

QM2 ENGINE ROOM INFORMATION

Diesel Engines:	4x 16-cylinder Wartsila 'V engines'
Gas Turbines:	2x General Electric LM2500+ gas turbines
Power:	157,000hp
Propulsion System:	4 pods of 20 MW each. 2 fixed and 2 azimuth
Bow Thrusters:	3x 3.2 MW thrusters
Stabilisers:	4x VM Series Stabilisers at 70 tons each
Anchors:	3x 23-ton anchors – 2 working and 1 spare
Anchor Chains:	3 anchor chains; collectively 843 yards (771m) long

◄ QE2 *sails alongside* QM2. *(Courtesy of Marc-Anthoine Bombail & Ocean Books)*

▼ *(Courtesy of Marc-Anthoine Bombail & Ocean Books)*

➤ *QM2 departs Cape Town on her maiden visit. (Jan Frame)*

➤➤ *QM2 by night. (Andrew Sassoli-Walker)*

INTERESTING FACTS

QM2 *has:*

- 2,500km of electric cable
- 500km (310 miles) of ducts, mains and pipes
- 2,000 bathrooms
- 80,000 lighting points
- 250,000m² (280,000 yards²) of fitted carpets
- 120,000m² (144,000 yards²) of insulating material
- 3,200m² (3,800 yards²) of galleys
- 3,000 telephones
- 8,800 loudspeakers
- 5,000 stairs
- 5,000 fire detectors
- 1,100 fire doors
- 8,350 automatic extinguishers

QM2 EXCLUSIVES

At the time of her introduction into service, QM2 *made history by having:*

- Over £3.5 million of artwork on board
- The first planetarium at sea, with virtual reality rides through the galaxies
- A cultural academy operated by the University of Oxford
- The first suites with private lift access
- The first Canyon Ranch Spaclub at sea
- The first Veuve Clicquot Champagne Bar at sea
- The largest library at sea (with 8,000 hardbacks, 500 paperbacks, 200 audio books and 100 CD ROMs)
- The largest ballroom with the largest dance floor at sea (measuring 7.5m by 13m)

- Workshops and master classes are performed by RADA (Royal Academy of Dramatic Arts)
- The longest jogging track at sea
- The largest and most extensive wine cellar at sea
- The '*Queen Mary 2*' signs near the funnel are the largest illuminated ship's name signs in maritime history

APPENDIX 2

GLOSSARY OF NAUTICAL (AND *QM2*) TERMS

Abeam	Off the side of the ship, at a 90° angle to its length
Aft	Near or towards the back of the ship
Amidships	Towards the middle of the ship
Azimuth Pod	A propeller pod that can be rotated in any horizontal direction
Blue Riband	Award presented for the fastest North Atlantic crossing
Boston Cup	Created to acknowledge the maiden arrival of *Britannia* in Boston. Now displayed aboard *QM2*
Bow	The forward-most part of a ship
Bow Thrusters	Propeller tubes that run through the width of the ship (at the bow) to help manoeuvrability
Bridge	Navigational command centre of the ship
Colours	The national flag or emblem flown by the ship
Draft	Depth of water measured from the surface of the water to the ship's keel
Forward	Near or towards the front of the ship
Hove to	When the ship is at open sea and not moving
Hull	The body of the vessel that stretches from the keel to the superstructure (*QM2*'s is painted black)

QM2's funnel.
(Authors' Collection)

Keel	The lowest point of a vessel
Knot	1 nautical mile per hour (1 nautical mile = 1,852 metres or 1.15 statute miles)
Leeward	The direction away from the wind
Maritime Quest	A self-guided tour aboard *QM2* that enlightens passengers of Cunard's unique history
Ocean Liner	A ship that undertakes a scheduled ocean service from point A to point B
Pitch	The alternate rise and fall of the ship which may be evident when at sea
Pods	Like giant outboard motors – the pods hang under *QM2* and provide propulsion, replacing the traditional propeller shafts
Port	The left side of the ship when facing forward
Starboard	The right side of the ship when facing forward
Stern	The rearmost part of a vessel
Superstructure	The body of the ship above the main deck or hull (*QM2*'s is painted white)
Tender	A small vessel (sometimes a lifeboat) used to transport passengers from ship to shore
Wake	The trail of disturbed water left behind the ship when it is moving
Windward	Direction the wind is blowing

Captain Nick Bates is well known for his sense of humour: 'According to my friend Patrick O'Shaughnessy there are three essential rules before you can become a successful captain on a cruise liner. Unfortunately nobody knows what they are!'

QM2 BUILDING TIMELINE

8 June 1998

Project *Queen Mary* announced just one week after Carnival Corporation completes its purchase of Cunard Line. Plans to undertake the design and development of a new class of transatlantic liner unveiled.

8 November 1999

Cunard announces that the general arrangement plans for the new liner completed. Project *Queen Mary* to be the largest passenger ship ever built. First image released.

Cunard's Norwegian shareholders were informed by Larry Pimentel, the Carnival Corporation appointed President and Chief Operating Officer of Cunard (at the time), who said, 'The project will lead to development of the heaviest liner ever built – the epitome of elegance, style and grace.'

10 March 2000

Letter of Intent signed with Chantiers de l'Atlantique shipyard in France for the US$700 million *Queen Mary 2*.

The magnificent QM2 at sea. (Matt Welsby)

6 November 2000

Formal contract signed in Paris by Micky Arison (Chairman and CEO of Carnival Corporation) and Patrick Boissier (President ALSTOM Marine and Chairman and CEO of Chantiers de l'Atlantique). At the same time in London the interior design for *QM2* is unveiled. *QM2* will be the largest, longest, tallest, widest and most expensive passenger ship ever.

January 2001

'Patron's Preview' programme launched granting passengers who sail on board *QE2* and *Caronia* in 2001 an exclusive month-long preview of *QM2*'s maiden season.

February 2001

Tank tests of *QM2* model successfully completed.

November 2001

Cunard announces Canyon Ranch will operate the health spa.

16 January 2002

Pamela Conover, Cunard's President and Chief Operating Officer, presses the button to cut the first sheet of steel for *QM2*.

March 2001

73 per cent of steel material ordered. Two panels (out of 580) completed.
6 per cent of the steel cut.

April 2001

82 per cent of steel material ordered. Six panels (out of 580) completed.
11 per cent of the steel cut.

May 2001

90 per cent of steel material ordered. Seven panels (out of 580) completed.
15 per cent of the steel cut (equating to 5,200 tons).

4 July 2002

Keel-laying ceremony takes place.

13 June 2003

Announcement made that Veuve Clicquot Ponsardin have agreed to create a champagne bar aboard *QM2*.

22 December 2003

QM2 is handed over to Cunard Line.

8 January 2004

QM2 is officially named by HM Queen Elizabeth II.

12 January 2004

QM2 departs on her maiden voyage from Southampton.

On the deck of QM2.
(Authors' Collection)

QM2 MILESTONES

22 December 2003

QM2 is handed over to Cunard Line.

8 January 2004

QM2 is officially named by HM Queen Elizabeth II.

12 January 2004

QM2 departs on her maiden voyage from Southampton.

26 January 2004

Completes maiden voyage.

17 February 2004

First equator crossing takes place.

21 February 2004

QM2 makes historic maiden call to Rio de Janeiro for Carnaval.

➤ *The beautiful* QM2. *(Matt Welsby)*

13 April 2004

QM2 featured on Royal Mail first-class stamp.

25 April 2004

QM2 and *QE2* set off for their first tandem transatlantic crossing.

1 May 2004

QM2 becomes flagship of the Cunard fleet, taking the title from Cunard's longest serving flagship, *QE2*.

17 May 2004

First ever *QM2* helicopter medical evacuation was carried out 89 miles west-south-west of Cape Hatteras.

19 July 2004

QM2 arrives in Hamburg, attracting over 1 million spectators.

11–30 August 2004

QM2 in Piraeus to serve as a floating hotel during the Athens Olympic Games.

11 July 2005

QM2 carries first signed copy of *Harry Potter and the Half Blood Prince* to the USA for book launch event.

23 February 2006

QM2 and *Queen Mary* meet in Longbeach, California.

12 May 2006

MS *Freedom of the Seas* is christened in New York Harbour. At 154,407 tons, she eclipses *QM2* as the world's largest passenger ship.

November 2006

Sprinklers are retrofitted to cabin balconies due to changes in safety requirements after the *Star Princess* fire. The work is completed at Blom & Voss, Hamburg. *QM2*'s bridge wings are extended to improve visibility.

4 February 2007

Maiden arrival in San Francisco attracts thousands of onlookers.

13 February 2007

Maiden call at Pago Pago involves tricky manoeuvre under cable car wiring.

20 February 2007

QM2 makes first call to Sydney. Later that day *QE2* joins her in Sydney with an estimated 2 million spectators bringing the city to a standstill.

13 January 2008

QM2 meets *QE2* and *Queen Victoria* in New York Harbour.

22 April 2008

QM2 arrives in Southampton to rendezvous with *QE2* and *Queen Victoria*, the last time all three Cunard Queens are seen together.

October 2008

QM2 and *QE2* undertake historic tandem transatlantic crossings to mark *QE2*'s retirement.

21 March 2009

QM2 meets *QE2* in Dubai.

October 2009

During round-Britain voyage, *QM2* makes maiden visits to Greenock (Scotland) and Liverpool (England), drawing large crowds.

27 January 2010

QM2 meets *QE2* in Dubai for a second time.

11 March 2010

QM2 makes maiden call in Adelaide.

14 March 2010

QM2 makes maiden call in Fremantle during 2010 world cruise.

20 March 2010

QM2 makes inaugural calls in Mauritius.

23 March 2010

QM2 makes maiden call in Durban, South Africa.

25–26 March 2010

QM2 overnights on her first visit to Cape Town.

April 2010

Due to the grounding of flights caused by Eyjafjallajokull Volcano in Iceland, *QM2* has a waiting list of over 1,000 passengers.

May 2010

QM2 returns to Manhattan due to damage of Brooklyn's Terminal.

25 June 2010

Princess Anne, HRH The Princess Royal, visits *QM2* in Southampton.

2 July 2010

QM2 hosts launch of Bill Miller's *Mr Ocean Liner* DVD with special luncheon aboard during her call in New York.

February 2012

QM2 to undertake circumnavigation of Australia.

QM2'S FAMOUS FACES

Since HM Queen Elizabeth II named *QM2* on 8 January 2004, the ship has hosted:

EMINENT PEOPLE WHO HAVE TRAVELLED

Prince Mubarek of Kuwait
Archbishop Desmond Tutu

WELL-KNOWN JOURNALISTS AND AUTHORS WHO HAVE TRAVELLED

Cindy Adams
Kate Adie
Jennie Bond
James Bardy
Tina Brown
Jean Chatzky
Mary Higgins Clark
Craig Doyle
Peter Greensberg
Sir Harold Evans
Sir Simon Jenkins
Sue MacGregor
John Maxtone-Graham
Richard Quest
Simon Schama
Lara Spencer
Graham Taylor
Terry Waite
William H. Miller Jr
Professor Lord Robert Winston

MUSICIANS WHO HAVE TRAVELLED

Dame Shirley Bassey
Opera Babes
Ruben Studdard
Carly Simon
Rod Stewart

STARS OF STAGE AND SCREEN WHO HAVE TRAVELLED

Karen Allen
John Cleese
Richard Dreyfuss
Carries Fisher
Richard Johnson
Patricia Neal
Des O'Connor
Paul Rudd
Jane Seymour

CULINARY ICONS WHO HAVE TRAVELLED

Daniel Boulud
Philip Cooper
Todd English
Laurent Gras
Jeffrey Vigilla

EMINENT PEOPLE WHO HAVE VISITED BUT NOT TRAVELLED

HM the Queen
HRH the Duke of Edinburgh
HRH The Princess Royal
Princess Michael of Kent
Queen Sofia of Spain
Rt Hon. Tony Blair
Mayor Michael Bloomberg
President George Bush Snr
Senator Hilary Clinton
President Jacques Chirac
Rt Hon. Alastair Darling, MP
Rt Hon. John Prescott, MP
Donald Trump

WELL-KNOWN JOURNALISTS AND AUTHORS WHO HAVE VISITED BUT NOT TRAVELLED

Dame Beryl Bainbridge
Steven Cojocaru 'Cojo'
Katie Couric
Frederick Forsyth
Mary Hart
Lester Holt
Karenna Ore-Schiff
Deborah Roberts
Al Roker
Leslie Thomas
Barbara Walters

MUSICIANS WHO HAVE VISITED BUT NOT TRAVELLED

Lesley Garrett
Jon Bon Jovi
Shawn Carter 'Jay Z'
Peter Cincotti
Harry Connick Jnr
Beyonce Knowles
Heather Small

STARS OF STAGE AND SCREEN WHO HAVE VISITED BUT NOT TRAVELLED

Ted Allen
Antonio Banderas
Michael Buerk
Glenn Close
Danny deVito
Kyan Douglas
Christine Ebersole
Dakota Fanning
Thom Filicia
Whoopi Goldberg
Melanie Griffith
Heather Headley
Elizabeth Hurley
Deborra-Lee Jackson
Hugh Jackman
Star Jones
Carson Kressley
Nathan Lane
Jonathan Lithgow
George Lucas
Brian Stokes Mitchell
The Muppets
Rea Perlman
Esther Rantzen
Lela Rochon
Jai Rodriguez
Sir Jimmy Savile OBE
Tony Sirico
Steve Tyrell

CULINARY ICONS WHO HAVE VISITED BUT NOT TRAVELLED

Valentino

Braynard, F.O. and Miller, W.H. (1991) *Picture History of the Cunard Line*, Dover, United Kingdom

Cowell, A. (2000) 'Belfast Shipyard Loses Bid to Build Queen Mary 2, and Many Jobs' in *New York Times*, 11 March 2000

Frame, C. and Cross, R. (2009) *The QE2 Story*, The History Press, United Kingdom

Frame, C. and Cross, R. (2009) *QM2: A Photographic Journey,* The History Press, United Kingdom

Gandy, M. (1982) 'The Britannia Cup' in *The Antiques Magazine*, July 1982 edn, Vol. 122, pp.156–8

Grant, R.G. (2007) *Flight: The Complete History*, Dorling Kindersley Ltd, United Kingdom

Maxtone-Graham, J. and Lloyd, H. (2004) *Queen Mary 2: The Greatest Ocean Liner of Our Time*, Bulfinch, United Kingdom

Miller, W.H. (2001) *Picture History of British Ocean Liners: 1900 to the Present*, Dover, United Kingdom

Miller, W.H. (1995) *Pictorial Encyclopaedia of Ocean Liners 1860–1994*, Dover, United Kingdom

Plisson, P. (2004) *Queen Mary 2: The Birth of a Legend*, Harry N. Abrams, United Kingdom

Cunard Line (2009) On-board Promotional Material (Various Versions)

Cunard Line (2009) *Queen Mary 2: Technical and Bridge Facts* (Various Versions)

Personal Conversations:

Commodore R.W. Warwick
Commodore B. Warner
Captain N. Bates
Captain C. Rynd
Chief Engineer B. Wattling
Food & Beverage Manager S. Engl
Marine Supervisor A. Gould
First Officer D. Hudson
First Officer J. Griffiths
Second Officer S. Roje
Third Officer H. Johnson
Third Officer C. Wood
Deck Cadet H. Morley
Captain's Secretary C. Hansen
Technical Secretary E. Wealleans
Cunard Line PR Executive M. Gallagher

Websites:

Chris' Cunard Page: http://www.chriscunard.com/ – includes profiles, history, facts and photographs of the Cunard fleet, past and present.

Sam Warwick's *QM2* Homepage: http://www.QM2.org.uk/ – excellent information including the ship's itinerary.

Cunard's Official UK Homepage: http://www.cunard.co.uk/

Further Reading:

Maritime books by Chris Frame & Rachelle Cross and The History Press:

QM2: A Photographic Journey

If you are interested in reading more about QM2*'s illustrious predecessor,* QE2*:*

QE2: A Photographic Journey
The QE2 Story

If you are interested in Cunard's Queen Victoria*:*

Queen Victoria: A Photographic Journey

The funnel houses the original Queen Mary *whistle. (Authors' Collection)*

➤ QM2's *bow with* QE2 *in the foreground during their 2009 meeting in Dubai. (Jan Frame)*